PLANTS!
HOW DO THEY GROW?

Annabel Griffin · Illustrated by Tjarda Borsboom

Copyright © 2024 Hungry Tomato Ltd

First published in 2024 by Hungry Tomato Ltd
F15, Old Bakery Studios, Blewetts Wharf, Malpas Road, Truro, Cornwall,
TR1 1QH, UK.

No part of this publication may be reproduced, stored in a retrieval system, or transmitted in any form or by any means, electronic, mechanical, photocopying, recording, or otherwise, without prior written permission of the copyright owner.

A CIP catalogue record for this book is available from the British Library.

ISBN 9781916598898

Printed in China

Discover more at
www.hungrytomato.com

Contents

What Is a Plant?	4	Amazing Bugs	16
What Do Plants Need?	6	Let's Grow Tomatoes	18
How Do Plants Grow?	8	Did You Know?	20
How Do Bees Help Plants?	10	Match Up the Pairs	22
What Do Seeds Look Like?	12	Glossary	24
How Do Plants Get Around?	14		

Words in **BOLD** can be found in the glossary.

Picture Credits

Abbreviations: m-middle, t-top, l-left, r-right, bg-background.

Shutterstock: Estragon 19br; Kazakova Maryia 18-19bg; Likar 23mr; Mazur Travel 23br; Nailia Schwarz 23ml; Photoongraphy 23tl; Skrypnykov Dmytro 23tr; Surked 23bl; Will Pedro 21tl; xpixel 20tl.

Every effort has been made to trace the copyright holders, and we apologise in advance for any unintentional omissions. We would be pleased to insert the appropriate

What Is a Plant?

Plants are living things that can be found almost everywhere on Earth! There are over 300,000 different types of plants on our planet. How many can you name?

Plants come in all sorts of shapes and sizes, but most of them have the same three parts: stem, roots, and leaves.

Leaves
Leaves are very important. They help the plant make its own food, to give it energy and help it grow.

Stem
A plant's stem grows above the ground and gives support. It acts as a drinking straw for the plant, carrying water and **nutrients** from the roots to different parts of the plant.

Roots
Roots are usually hidden underground. They help to hold the plant in place, like an anchor. They also take up water and nutrients from the soil that the plant needs to grow.

What Do Plants Need?

Plants can't grow without a few very important things: sunlight, air, water, and nutrients.

Sun
Plants take light from the sun and turn it into food, which gives them energy to grow.

Air and water
Without enough air and water, plants will quickly shrivel up and die.

Nutrients
The roots of a plant take up water and nutrients (food) from the soil.

Space
Some plants need space away from other plants to avoid **competing** for nutrients in the soil.

How Do Plants Grow?

Most plants grow from seeds. Even a gigantic tree starts its life as a tiny seed. How does it all begin?

Seeds

When a seed is planted and given water, it will begin to grow into a plant. This is called **germination**.

Shoot

1.
A root starts to grow out from the seed.

2.
A shoot grows up from the seed towards the soil's surface.

3.
The roots continue to grow as the shoot becomes a stem and starts to grow leaves.

flower bud

4. The plant grows bigger and stronger, creating more leaves and forming a flower bud.

5. The plant reaches its full size and its flower opens.

Let's Grow a Seed

You will need:
- Sunflower seeds
- Large plant pot
- Peat-free potting compost

seeds

1. Read the seed packet to find the best time for planting.

2. Fill a large flowerpot with potting compost. Push a seed about 1cm (½ inch) down into the compost and cover it fully.

3. Place it in a sunny spot and water it regularly, so it doesn't dry out.

Flower

How Do Bees Help Plants?

Bees are very helpful friends to plants. They carry **pollen** from one flower to another. This is called **pollination**. Plants need pollen from other flowers to make new seeds.

1. Bees are attracted to a flower's bright petals and lovely scent.

2. Flowers contain nectar, a sweet liquid that bees collect to make honey.

You look lovely!

Yummy nectar!

3.
As a bee collects the nectar, it becomes covered in pollen, which is carried to the next flower it visits.

What Do Seeds Look Like?

Seeds come in lots of different shapes and sizes. Some seeds are nice to eat, and some are hidden inside fruit. Have you seen any of these seeds before?

Poppy
Once pollinated, a poppy will lose its petals and grow a **seedpod**, full of tiny black seeds.

The seedpod has holes, like a saltshaker, so seeds can blow out in the wind.

Sunflower
Sunflower seeds grow in the middle of the flower.

Wheat
Wheat seeds are used to make flour. They grow at the top of each stalk.

Amazing Bugs

Did you spot the creepy-crawlies hidden throughout this book? Creepy-crawlies are an important part of keeping plants healthy!

Buzzing bumblebees
Bees are one of the most important bugs. We rely on them to spread pollen, which helps flowers, fruits, and vegetables to grow.

Daring dragonflies
These prehistoric bugs have existed for over 300 million years! Dragonflies love ponds and wildflower meadows, where they hunt plant-eating bugs.

Wiggly worms
Worms are invertebrates, which means they don't have any bones! They wiggle through soil, eating dead plants and leaving behind nutrients for plants to soak up.

Beautiful butterflies

Butterflies have brightly patterned wings, making them the prettiest bug of all. Like bees, they drink nectar from flowers and pollinate gardens and meadows, helping new plants to grow.

Pesky flies

It's true that some flies buzz around, eating plants, fruits, and vegetables. However, many flies actually help plants grow by pollinating them, just like bees do!

Brilliant beetles

Beetles can be all sorts of shapes and sizes. They are great because they hunt and eat bugs, like slugs and snails, that destroy plants.

Let's Grow Tomatoes

See the magic of plants in action by growing your very own tomatoes! You can grow them outside or in a greenhouse, as long as they get plenty of sun. Ask an adult to help you choose a good place for your plants before starting.

You will need:

- Packet of tomato seeds
- Soil/peat-free potting compost
- Mini plant pots/seed trays
- Medium to large pots
- Tomato feed
- Plastic wrap
- Long sticks (to hold the plant upright)
- String

Important tips:

- Always follow the instructions on the seed packet to know when to **sow**, plant, and **harvest** your tomatoes.

- Don't worry if not all your seeds grow – that's very normal. That's why it's important to plant more than 1 seed!

- If any of your plants struggle to stand up on their own, tie the stem to a stick to help the plant out.

1.
Fill your mini pots/seed tray with compost to about 1cm (½ inch) from the top.

2.
Space out 1-2 seeds on the surface, then cover with a thin layer of soil. Lightly water, cover with plastic wrap, and place somewhere warm and light.

3.
In 5-10 days, the seeds should start to sprout. Once you see the plant above the soil, take off the plastic wrap. Keep the soil moist by watering it regularly.

Watch out for hungry bugs that will eat your plants!

4.
When the seedlings are 7.5cm (3 inches), move them into the medium pots. Hold them by the leaves, not the stems. Water every day.

5.
When flowers start appearing, carefully move your plants into the biggest pots. Now they have space to grow.

6.
Now, put the plants outside or in a greenhouse. Give them tomato feed every week and keep watering them.

Green
These tomatoes aren't ripe yet; they taste quite bitter!

Orange
These tomatoes are nearly ready to eat!

7.
The flowers should turn into tomatoes! Be patient while they grow; it can take a few months from sowing the seeds to harvesting the fruit, but it will be worth the wait!

Red
When the tomatoes change to red, they are ready to eat. Enjoy!

Did You Know?

Plants are pretty amazing! Every living creature needs plants to survive; the world wouldn't be the way it is today if we didn't have them. Did you know these amazing facts about plants?

Scientists can work out the **AGE** of a tree by looking at the number of rings in its trunk!

One strawberry has around **200** seeds. That's a lot!

Have you ever wondered why an apple can float in water? It's because they are made up of **25%** air!

Bristlecone pine tree

The oldest living plant on Earth is a bristlecone pine tree in California (USA), which is almost **5,000** years old!

A single bee can pollinate over **1,000** flowering plants in one day!

Around **80%** of all flowering plants rely on insects, such as bees and butterflies, for pollination. Where would we be without them?

Match Up the Pairs

Can you match up the fact boxes (below) with the correct plant (right)? Flip back through the book if you need a hint!

1.
My sweet, yellow seeds are called kernels. They are yummy to eat.

2.
I'm a pretty flower. My tiny, black seeds grow in a seedpod after I lose my petals.

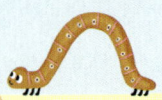

3.
I'm a big, tall flower that's yellow, like the sun.

4.
I'm a tree and I like to grow near water, so that my seeds can be spread by the current.

5.
I'm a tree which grows tasty green or red fruit.

6.
I'm a plant that grows sweet fruit whose seeds grow on the outside.

Corn

Sunflower

Poppy

Strawberry plant

Coconut tree

Apple tree

Have you matched them all?
Answers can be found on page 24.

Glossary

Competing – (verb) going against one another to gain or win something.

Current – the continuous movement of a body of water, such as a river or ocean.

Germination – the process when a seed begins to sprout roots and shoots.

Harvest – (verb) the process of gathering in crops when they are ripe and ready to be gathered.

Nutrients – substances or ingredients that plants and animals need to live and grow.

Pollen – a dusty powder made by some plants. It is used within pollination (see below) to produce new seeds.

Pollination – when pollen (see above) is moved from one plant to another – often by an insect – so that the plants can make new seeds.

Seedpod – a pouch or case produced by some plants to hold their seeds.

Sow – (verb) to plant or scatter seeds into soil.

Trunk – the large woody stem of a tree, where the branches grow from.

Answers to Match Up the Pairs

Answers: 1. Corn, 2. Poppy, 3. Sunflower, 4. Coconut tree, 5. Apple tree, 6. Strawberry plant.